SELF-SUFFICIENCY FARMING FOR BEGINNERS

By

DR. ANITA BRAINERD

CHAPTER ONE

INTRODUCTION TO SELF-SUFFICIENCY FARMING

1. What is self-sufficiency farming?

Self-sufficiency farming refers to a farming practice where individuals or communities aim to produce the majority of their own food and resources, minimizing reliance on external sources. This approach involves cultivating a diverse range of crops, raising livestock, and possibly incorporating sustainable farming methods. The goal is to meet one's nutritional needs and reduce dependence on commercial agriculture and food supply chains.

Practitioners of self-sufficiency farming often prioritize organic and regenerative farming techniques, emphasizing soil health and biodiversity. Additionally, they may employ methods such as permaculture, agroforestry, and water conservation to create resilient and sustainable agricultural systems.

By fostering independence and reducing external inputs, self-sufficiency farming aims to enhance food security, promote environmental sustainability, and build resilient communities. This approach reflects a desire to create a more sustainable and localized food production system, decreasing the ecological footprint associated with conventional agriculture and fostering a deeper connection between individuals and the food they consume.

2. Benefits and challenges of self-sufficiency farming.

➢ **Benefits:**

- Food Security: Self-sufficiency farming ensures a consistent and reliable source of food for the household or community, reducing dependence on external markets and fluctuations in prices.

- Cost Savings: By producing their own food and resources, individuals or communities can save money that would otherwise be spent on purchasing these items. This can lead to increased financial stability.

- Environmental Sustainability: Self-sufficiency farming often promotes sustainable agricultural practices, such as organic farming, crop rotation, and agroforestry, which can be beneficial for the environment.

- Community Building: Self-sufficiency farming encourages communal collaboration and mutual support within a community. Sharing resources, knowledge, and labor can strengthen social ties.

- Independence: Being self-sufficient reduces dependence on external entities, making individuals or communities more resilient to economic, political, or environmental shocks.

- Local Economy Support: Self-sufficiency farming can contribute to the development of local economies by keeping resources within the community and fostering entrepreneurship.

➤ **Challenges:**

- Limited Diversity: Self-sufficiency farming may lead to limited crop and livestock diversity, potentially resulting in nutritional deficiencies and increased vulnerability to pests and diseases.

- Time and Labor Intensive: Maintaining a self-sufficient farm requires significant time and labor. This can be challenging for individuals or communities with limited resources or other obligations.

- Lack of Specialization: Self-sufficiency farming often involves a broad range of activities, making it difficult to specialize in specific crops or livestock. This can impact overall efficiency and productivity.

- Weather Dependency: Self-sufficient farmers are highly dependent on weather conditions. Unfavorable weather patterns, such as droughts or floods, can have a significant impact on crop yields.

- Infrastructure and Technology Challenges: Lack of access to modern farming equipment, technologies, and infrastructure can hinder productivity and make it challenging to compete with more industrialized farming methods.

- Market Exclusion: Self-sufficiency farming may limit access to broader markets, reducing opportunities for income generation and economic growth.

3. Understanding the principles of self-sufficiency.

- **Independence:** Self-sufficiency promotes independence by reducing dependence on others for basic needs such as food, shelter, and clothing.

- **Resourcefulness:** It encourages individuals to be resourceful and creative in finding solutions to their needs, often leading to innovation and efficiency.

- **Sustainability:** Self-sufficiency emphasizes sustainable practices that minimize environmental impact and ensure the long-term viability of resources.

- Skill Development: Pursuing self-sufficiency involves acquiring diverse skills, from gardening and food preservation to carpentry and basic healthcare, fostering personal growth and resilience.

- **Community Engagement:** While self-sufficiency often conjures images of isolation, it can also strengthen community ties through shared knowledge, resources, and mutual support networks.

- Economic Resilience: By reducing reliance on external markets and systems, self-sufficiency can enhance economic resilience, particularly in times of crisis or uncertainty.

- **Cultural Preservation:** Many cultures value self-sufficiency as a way to preserve traditional knowledge and practices passed down through generations.

- **Flexibility:** Self-sufficiency doesn't necessitate complete isolation but rather encourages flexibility in utilizing both self-produced and externally sourced resources as needed.

CHAPTER TWO

PLANNING YOUR SELF-SUFFICIENCY FARM

1. Assessing your land and resources.

➢ **Land Evaluation:**

- Soil Quality: Conduct soil tests to determine its composition, fertility, and pH levels. This information helps in choosing suitable crops and deciding on necessary soil amendments.

- Topography: Understand the landscape, slopes, and drainage patterns to plan for water management, erosion prevention, and suitable crop placement.

- Climate: Consider the local climate, including temperature, rainfall, and seasonal variations, to select crops and animals that thrive in your region.

➢ **Water Resources:**

- Water Availability: Assess the availability of water sources such as wells, rivers, or ponds. Plan for irrigation systems and water storage to support crops during dry periods.

- Quality Testing: Test the quality of water sources to ensure it is suitable for irrigation and animal consumption.

➢ **Plant and Animal Selection:**

- Crop Selection: Choose crops that are well-suited to your climate, soil, and water availability. Consider diversifying your crops to enhance resilience and self-sufficiency.

- Livestock: If you plan to raise animals, evaluate the space, feed availability, and infrastructure needed. Select livestock that aligns with your goals, considering factors like climate adaptation, local market demand, and sustainability.

➢ **Infrastructure and Facilities:**

- Buildings: Assess existing structures or plan for construction based on the needs of your farm, such as storage facilities, barns, and shelters for livestock.

- Fencing: Evaluate the need for fencing to protect crops and contain animals. Consider natural barriers and potential risks from wildlife.

> **Energy Sources:**

- Renewable Energy: Explore options for incorporating renewable energy sources such as solar panels or wind turbines to meet the farm's energy needs. This can enhance sustainability and reduce reliance on external energy.

> **Waste Management:**

- Composting: Implement composting systems for organic waste to create nutrient-rich soil amendments.

- Livestock Waste: Plan for proper management of animal waste to avoid environmental issues and utilize it for compost or fertilization.

> **Equipment and Tools:**

Inventory: Assess the tools and equipment available on the farm. Ensure they are well-maintained and suitable for the tasks at hand. Identify any gaps in your equipment inventory.

> **Financial Planning:**

- Budgeting: Create a budget outlining your expenses and potential income. Consider initial investment costs, ongoing operational expenses, and potential revenue streams.

➤ **Skill Assessment:**

- Training: Evaluate your own skills and those of anyone involved in the farm. Consider acquiring additional knowledge through training programs or workshops.

➤ **Emergency Preparedness:**

- Contingency Plans: Develop contingency plans for unforeseen events such as natural disasters, crop failures, or market fluctuations. This includes having emergency supplies and evacuation plans if necessary.

2. Setting goals for your farm

- **Food Production:** Determine the types and quantities of food you want to produce on your farm, considering nutritional needs, dietary preferences, and seasonal availability. This might include fruits, vegetables, grains, herbs, and even livestock for meat, dairy, and eggs.

- **Diversification:** Aim to cultivate a diverse range of crops and livestock to ensure resilience against pests, diseases, and market fluctuations. Diversity can also enhance soil health and ecosystem balance.

- **Self-Sufficiency:** Set a target for the percentage of your family's food needs that you aim to meet through on-farm production. Gradually work towards reducing reliance on external sources for food and inputs like fertilizers and pesticides.

- **Sustainable Practices:** Commit to implementing sustainable farming methods that minimize environmental impact and promote long-term soil fertility and biodiversity. This might include organic farming, permaculture techniques, water conservation, and renewable energy use.

- **Soil Health:** Focus on improving and maintaining soil health through practices such as crop rotation, cover cropping, composting, and minimal tillage. Set goals for enhancing soil structure, fertility, and microbial activity.

- **Water Management:** Develop strategies to efficiently manage water resources on your farm, including rainwater harvesting, irrigation system upgrades, and soil moisture conservation techniques.

- **Energy Independence:** Work towards reducing dependence on external energy sources by investing in renewable energy technologies like solar panels, wind turbines, or biomass digesters.

- **Waste Reduction:** Aim to minimize waste generation and implement recycling and composting systems to utilize organic matter efficiently and reduce environmental pollution.

- **Community Engagement:** Consider goals related to community outreach and education, such as hosting workshops, farm tours, or volunteer opportunities to promote self-sufficiency, sustainable agriculture, and local food systems.

- **Financial Viability:** Set targets for achieving financial sustainability through diversified income streams, cost-effective resource management, and potentially value-added products like artisanal foods, crafts, or agritourism experiences.

- **Personal Well-being:** Don't forget to prioritize your own well-being and quality of life as a farmer. Set goals for maintaining work-life balance, managing stress, and fostering a sense of fulfillment and connection to the land and community.

3. Designing your farm layout for efficiency and sustainability

- **Assess Your Needs and Goals:** Determine what you want to achieve with your self-sufficiency farm. Consider factors such as food production for personal consumption, income generation through surplus produce sales, livestock raising, and self-reliance on energy and water.

- **Site Analysis:** Evaluate your land for factors like topography, soil type, drainage patterns, sunlight exposure, and prevailing winds. This analysis will help you understand the limitations and opportunities of your site.

- **Zoning:** Divide your land into zones based on factors like soil fertility, water availability, and microclimates. Allocate different areas for various purposes such as vegetable gardens, orchards, livestock pastures, and agroforestry systems.

- **Permaculture Principles:** Integrate permaculture principles into your farm design to maximize efficiency and sustainability. This includes practices like polyculture planting, companion planting, crop rotation, and using natural resources like mulch and compost.

- **Water Management:** Implement water conservation and management strategies such as rainwater harvesting, drip irrigation, swales, and contour farming to optimize water usage and minimize runoff.

- **Energy Efficiency:** Incorporate renewable energy sources such as solar panels, wind turbines, and biogas digesters to meet your farm's energy needs. Design energy-efficient infrastructure and equipment to reduce reliance on fossil fuels.

- **Crop Rotation and Companion Planting:** Plan crop rotations to improve soil health, minimize pests and diseases, and maximize yields. Utilize companion planting techniques to enhance biodiversity, attract beneficial insects, and suppress weeds.

- **Livestock Integration:** Integrate livestock into your farm layout in a way that complements other farming activities. Use rotational grazing systems to prevent overgrazing and soil erosion while improving pasture productivity.

- **Wildlife Habitat Preservation:** Designate areas of your farm for wildlife habitat preservation to promote biodiversity and ecosystem resilience. Create hedgerows, ponds, and native plantings to provide food, shelter, and nesting sites for beneficial wildlife species.

- **Infrastructure and Access:** Plan efficient infrastructure layout for easy access to different areas of your farm. Designate pathways, roads, and storage facilities strategically to minimize labor and transportation costs.

- **Record Keeping and Evaluation:** Keep detailed records of your farm activities, inputs, and outputs to track progress, identify areas for improvement, and make informed decisions for future planning.

- **Continuous Learning and Adaptation:** Stay informed about new technologies, techniques, and best practices in self-sufficiency farming. Continuously evaluate and adapt your farm layout and management strategies based on your experiences and changing circumstances.

CHAPTER THREE

BASIC FARMING PRACTICES

1. Soil health and fertility management.

- **Crop Rotation:** Implementing crop rotation helps prevent soil depletion by alternating plant families grown in specific areas each season. This practice minimizes nutrient depletion and reduces the buildup of pests and diseases.

- **Composting:** Recycling organic waste materials such as kitchen scraps, crop residues, and animal manure into compost enriches the soil with essential nutrients, improves soil structure, and enhances microbial activity, promoting overall soil health.

- **Cover Cropping:** Planting cover crops during fallow periods or in between cash crops helps prevent soil erosion, suppresses weed growth, and adds organic matter to the soil when they are incorporated or left as mulch, improving soil fertility and structure.

- **Mulching:** Applying organic mulches like straw, leaves, or grass clippings on the soil surface conserves moisture, regulates soil temperature, suppresses weeds, and gradually breaks down to enrich the soil with nutrients, enhancing fertility.

- **No-Till Farming:** Minimizing soil disturbance through no-till or reduced tillage practices preserves soil structure, organic matter, and beneficial soil organisms. This approach reduces erosion, promotes water infiltration, and enhances soil biodiversity, contributing to long-term soil health.

- **Crop Diversity:** Growing a diverse range of crops promotes biodiversity above and below ground, improving nutrient cycling, pest management, and overall soil resilience. Different crops have varied nutrient needs, reducing the risk of nutrient depletion.

- **Soil Testing and Amendments:** Regular soil testing helps assess nutrient levels and pH, guiding the application of targeted organic or mineral amendments to address deficiencies and maintain optimal soil fertility levels for crop growth.

- **Water Management:** Efficient water management practices such as drip irrigation or rainwater harvesting minimize soil erosion and nutrient leaching, ensuring that plants receive adequate moisture without compromising soil health.

2. Crop selection and rotation.

In self-sufficiency farming, crop selection and rotation are crucial practices that contribute to sustainable agricultural systems. The choice of crops should align with the local climate, soil conditions, and the farmer's nutritional needs. Diversifying crops helps mitigate risks associated with pests, diseases, and adverse weather conditions.

When selecting crops, consider those well-suited to the region's climate. This includes assessing temperature ranges, precipitation levels, and sunlight exposure. Additionally, soil composition plays a pivotal role, as certain crops thrive in specific soil types. A mix of vegetables, grains, fruits, and legumes can enhance nutritional diversity, ensuring a well-rounded diet.

Crop rotation is a strategic approach to maintaining soil health and fertility. By alternating crops in different seasons, farmers prevent the depletion of specific nutrients and reduce the risk of soil-borne diseases. For example, legumes, such as peas or beans, fix nitrogen in the soil, benefiting subsequent crops that require this essential nutrient.

A well-designed rotation plan considers the specific needs of each crop and optimizes resource utilization. It also helps break pest and disease cycles, as pathogens and insects often target specific plant families. The careful sequencing of crops can disrupt these cycles, reducing the reliance on chemical interventions.

In summary, successful self-sufficiency farming demands thoughtful crop selection and rotation. By adapting choices to local conditions and implementing a strategic rotation plan, farmers can optimize yields, enhance soil fertility, and foster a resilient and sustainable agricultural ecosystem on their small-scale farms.

3. Water management techniques.

- **Rainwater Harvesting:** Collecting rainwater through systems like rain barrels, swales, and ponds can provide a significant portion of water needs, especially during dry spells. This stored water can be used for irrigation, livestock, and household purposes.

- **Drip Irrigation:** Drip irrigation delivers water directly to the base of plants, minimizing evaporation and water waste compared to traditional overhead irrigation methods. It's efficient and can be set up with timers to optimize water usage.

- **Mulching:** Applying organic mulch like straw, wood chips, or compost to the soil surface helps retain moisture by reducing evaporation. Mulch also improves soil structure and fertility over time.

- **Permaculture Design:** Implementing permaculture principles such as planting drought-resistant crops, creating swales to capture and channel water, and designing landscapes to maximize water retention can enhance water efficiency in farming systems.

- **Greywater Recycling:** Reusing household greywater (from sinks, showers, and washing machines) for irrigation purposes after treatment can supplement water resources on the farm.

- **Soil Moisture Monitoring:** Utilizing soil moisture sensors helps farmers understand the water needs of their crops more accurately, allowing for precise irrigation scheduling and preventing overwatering.

- **Crop Rotation and Cover Crops:** Implementing crop rotation and cover cropping strategies helps maintain soil moisture levels, improve soil structure, and reduce water runoff by keeping the soil covered with vegetation year-round.

- **Contour Farming:** Contour farming involves planting crops along the natural contours of the land, which helps slow down water runoff, reduce erosion, and promote infiltration of water into the soil.

- **Water-efficient Crop Selection:** Choosing crop varieties that are adapted to local climatic conditions and require less water can significantly reduce overall water usage on the farm.

- **Integrated Water Management Systems:** Integrating various water management techniques into a holistic system tailored to the specific needs and characteristics of the farm can maximize water efficiency and resilience against droughts or water scarcity.

CHAPTER FOUR

LIVESTOCK AND ANIMAL HUSBANDRY

1. Raising chickens, goats, or other small livestock for meat, eggs, and dairy.

Raising chickens, goats, or other small livestock for meat, eggs, and dairy is a cornerstone of self-sufficiency farming. These animals provide a renewable source of protein and other essential nutrients while also offering additional benefits like pest control and soil enrichment.

Chickens are a popular choice due to their versatility. They require minimal space and are relatively low-maintenance, making them ideal for small-scale operations. In addition to meat and eggs, chickens produce valuable fertilizer for gardens and can help control insect populations.

Goats are another excellent option for self-sufficiency farming. They are efficient converters of roughage into milk and meat, making them valuable for dairy and meat production. Goats are also well-suited to small farms as they can thrive on a variety of vegetation and are generally hardy animals.

Other small livestock such as rabbits, ducks, and quail can also play a role in self-sufficiency farming. Rabbits are efficient producers of meat, while ducks offer both meat and eggs. Quail are well-suited for small spaces and can provide a steady supply of eggs.

In self-sufficiency farming, careful planning and management are essential to ensure the health and productivity of the animals. Providing appropriate housing, nutrition, and healthcare are crucial aspects of livestock management. Additionally, rotational grazing and sustainable land management practices can help maximize the productivity of the farm while preserving natural resources.

By raising chickens, goats, or other small livestock, self-sufficiency farmers can reduce their reliance on external food sources and create a more sustainable and resilient food system. These animals not only provide valuable products for the farm but also contribute to a healthier environment and a greater sense of self-reliance.

2. Integrating animals into your farm ecosystem.

Integrating animals into your self-sufficiency farming system can enhance the overall sustainability and productivity of your farm ecosystem. Livestock play a crucial role in providing valuable resources such as manure, which serves as a natural fertilizer, promoting soil health and fertility.

Chickens, for example, not only supply eggs but also help control pests by foraging for insects in the fields. Their droppings contribute to nutrient-rich compost, supporting plant growth. Similarly, goats and sheep can serve as efficient grass trimmers, reducing the need for mechanical mowing while supplying milk, meat, and wool.

Cattle, if space allows, contribute to a diversified farming system by providing meat and dairy products. Their grazing behavior also aids in managing vegetation, preventing overgrowth and promoting biodiversity.

In addition to tangible products, animals contribute to a balanced ecosystem by cycling nutrients through their waste and creating a symbiotic relationship with plants. Rotational grazing, where animals are moved through different sections of the farm, prevents overgrazing and allows vegetation to regenerate.

Integrating bees as pollinators enhances crop yields and promotes biodiversity. Bees play a vital role in fruit and vegetable production, ensuring a more abundant harvest.

In a self-sufficiency farm, careful planning and consideration of animal selection, rotation, and integration are essential. This holistic approach fosters a resilient and sustainable agricultural system, where each component complements the others, reducing reliance on external inputs and creating a harmonious, closed-loop ecosystem.

3. Sustainable animal husbandry practices.

- **Rotational Grazing:** Implementing rotational grazing allows pastures to rest and recover while preventing overgrazing. It also promotes soil health, biodiversity, and carbon sequestration.

- **Pasture Management:** Maintaining healthy pastures through proper soil management, diverse forage species, and minimal chemical inputs ensures sufficient nutrition for animals while reducing reliance on external resources.

- **Integrated Livestock and Crop Production:** Utilizing animal manure as fertilizer for crops closes nutrient cycles, reducing the need for synthetic fertilizers and improving soil fertility. Additionally, crop residues can be used as feed, minimizing waste.

- **Diversification of Livestock:** Raising a variety of livestock species promotes resilience against diseases and market fluctuations. It also allows for better utilization of resources, such as grazing lands and feed resources.

- **Heritage Breeds and Local Adaptations:** Breeding and raising heritage breeds and locally adapted livestock breeds are often better suited to the environment and require fewer inputs such as medication and specialized feed.

- **Water Management:** Implementing water-saving techniques such as rainwater harvesting, efficient irrigation systems, and proper watering schedules reduces water wastage and ensures adequate hydration for animals.

- **Natural Pest Control:** Encouraging natural predators and implementing integrated pest management practices minimizes the need for chemical pesticides and reduces environmental contamination.

- **Animal Welfare:** Providing appropriate housing, access to outdoor space, and veterinary care ensures the well-being of animals, leading to healthier livestock and higher-quality products.

- **Energy Efficiency:** Implementing energy-efficient infrastructure and utilizing renewable energy sources such as solar power reduces the carbon footprint of animal husbandry operations.

- **Continuous Learning and Improvement:** Regularly seeking knowledge about sustainable practices, staying informed about advancements in the field, and adapting methods accordingly ensures continual improvement in sustainability efforts.

CHAPTER FIVE

FOOD PRESERVATION AND STORAGE

1. Techniques for preserving fruits and vegetables.

- **Canning:** This traditional method involves heating fruits or vegetables in jars to kill bacteria and seal them airtight. It preserves nutrients and flavors for long periods. Water bath canning is suitable for high-acid foods like tomatoes and fruits, while pressure canning is necessary for low-acid vegetables.

- **Freezing:** Freezing is a simple and effective way to preserve fruits and vegetables. It involves blanching them briefly in boiling water to stop enzyme activity, then quickly cooling and freezing. Use airtight containers or freezer bags to prevent freezer burn.

- **Drying/Dehydrating:**   Dehydrating   removes moisture from fruits and vegetables, inhibiting microbial growth and preserving nutrients. You can sun-dry produce or use a dehydrator for consistent results. Dried fruits and vegetables are compact and easy to store.

- **Pickling:** Pickling involves preserving produce in a solution of vinegar, salt, and spices. This acidic environment inhibits bacterial growth and adds flavor. Quick pickling is suitable for short-term storage, while traditional fermentation produces long-lasting pickles.

- **Root Cellaring:** Root cellars provide a cool, dark, and humid environment ideal for storing root vegetables like potatoes, carrots, and beets. Proper ventilation and temperature control are crucial to prevent spoilage.

- **Fermentation:** Fermenting vegetables like cabbage (for sauerkraut) or cucumbers (for pickles) involves natural bacteria breaking down sugars and producing lactic acid, which preserves the food. Fermented foods are not only preserved but also rich in probiotics, promoting gut health.

- **Preserving in Oil or Vinegar:** Immersing herbs or vegetables in oil or vinegar creates a hostile environment for bacteria. This method preserves flavors and can be used for ingredients like garlic cloves, peppers, or herbs.

- **Cold Storage:** Some fruits and vegetables, such as apples, pears, and squash, can be stored in a cool, dry place like a cellar or garage for several months, extending their shelf life without any preservation method.

2. Canning, drying, and fermenting food.

Self-sufficiency farming involves preserving food through canning, drying, and fermenting techniques to ensure a year-round supply of nutritious produce.

Canning, a popular preservation method, involves sealing food in airtight containers after heating to destroy bacteria. This extends the shelf life of fruits, vegetables, and even meats. Home canning allows farmers to store surplus harvests and enjoy the bounty during leaner months.

Drying is another efficient preservation technique. By removing moisture from fruits, vegetables, and herbs, farmers can prevent spoilage and retain nutritional value. Sun-drying and using dehydrators are common methods. Dried foods are lightweight, occupy less space, and are easy to store, making them ideal for long-term self-sufficiency.

Fermenting involves the natural breakdown of food by microorganisms, creating an environment that preserves and enhances flavors. This process not only extends shelf life but also boosts the nutritional content and promotes gut health. Fermented foods like sauerkraut, kimchi, and pickles are staples in a self-sufficiency pantry.

By mastering these preservation techniques, self-sufficiency farmers can minimize waste, reduce reliance on external food sources, and enjoy a diverse and nutritious diet year-round. Additionally, these practices contribute to sustainability by promoting local, seasonal eating and reducing the carbon footprint associated with long-distance food transportation.

In conclusion, canning, drying, and fermenting are integral components of self-sufficiency farming. These methods empower individuals to take control of their food supply, fostering resilience and sustainability within their communities.

3. Building root cellars and other storage facilities.

Root cellars and other storage facilities are indispensable components of self-sufficiency farming, ensuring the preservation of harvested produce for extended periods without reliance on external resources. These structures provide a cool, dark, and often humid environment ideal for storing root vegetables, fruits, and other perishables, effectively extending their shelf life and reducing waste.

Root cellars, traditionally dug into the earth or built partially underground, offer natural insulation against temperature fluctuations, maintaining a consistent cool temperature conducive to food preservation. Ventilation systems regulate humidity levels, preventing mold and decay while preserving the freshness of stored goods. Additionally, the darkness of a root cellar inhibits sprouting in root vegetables, further prolonging their storage life.

Beyond root cellars, self-sufficiency farmers may employ various storage facilities tailored to different types of produce. For instance, dry storage rooms with proper shelving and airflow are suitable for grains, seeds, and dried herbs. Cold storage rooms or refrigerated units enable the preservation of perishable items like dairy, eggs, and fresh vegetables.

Innovative approaches to storage, such as insulated containers or repurposed shipping containers, offer additional flexibility for small-scale farmers with limited space or resources. These containers can be modified with insulation, ventilation, and temperature control mechanisms to create efficient storage solutions for diverse agricultural products.

Implementing efficient storage facilities is essential for self-sufficiency farming, enabling farmers to maximize the yield of their harvests and minimize food waste. By investing in well-designed root cellars and storage structures, farmers can extend the availability of fresh, homegrown produce throughout the year, reducing dependency on external food sources and enhancing the sustainability of their operations.

CHAPTER SIX

BUILDING COMMUNITY AND SELF-RELIANCE

1. Engaging with local communities and sharing resources.

Engaging with local communities and sharing resources in self-sufficiency farming is essential for fostering resilience, sustainability, and mutual support. One key aspect is establishing community gardens or farms where individuals can collectively grow food and exchange knowledge and resources. These spaces serve as hubs for learning, collaboration, and building strong social connections.

Sharing resources such as seeds, tools, and expertise enables community members to overcome individual limitations and achieve greater productivity. Seed swaps, tool libraries, and skill-sharing workshops are effective ways to facilitate resource sharing within the community. Additionally, organizing work parties or volunteer days encourages collective action in tasks like planting, harvesting, and infrastructure maintenance.

Education plays a crucial role in promoting self-sufficiency farming within local communities. Offering workshops, demonstrations, and educational programs on topics like organic gardening, permaculture, and sustainable agriculture empowers individuals with the knowledge and skills needed to grow their own food. Moreover, creating opportunities for intergenerational learning allows for the transmission of traditional agricultural practices alongside modern techniques.

Engaging with local schools, community centers, and grassroots organizations can broaden the reach of self-sufficiency farming initiatives and attract diverse participants. Collaborating with government agencies and nonprofits can provide access to funding, technical support, and policy advocacy to strengthen community-based farming efforts.

Ultimately, building a resilient and self-sufficient local food system requires active participation and collaboration from community members, organizations, and stakeholders. By sharing resources, knowledge, and experiences, communities can create sustainable solutions to food insecurity, promote environmental stewardship, and enhance social cohesion.

2. Bartering and trading surplus produce.

Self-sufficiency farming is a practice that emphasizes the production of essential goods within a community or individual farm, minimizing dependence on external sources. Bartering and trading surplus produce play a crucial role in enhancing the sustainability and resilience of self-sufficiency farming.

In self-sufficiency farming, farmers prioritize growing a diverse range of crops and raising livestock to meet their basic needs. Surpluses often arise from this diversified production, creating an opportunity for bartering and trading with neighboring farmers or community members. This system fosters a symbiotic relationship where each party can exchange excess goods to fulfill their respective needs.

Bartering becomes a means of acquiring items that may not be efficiently produced on one's own farm. For instance, a farmer with surplus vegetables may trade with another farmer who has an abundance of eggs or dairy products. This exchange not only provides variety in the diet but also ensures a well-rounded nutritional intake.

Trading surplus produce also contributes to the local economy and community cohesion. It establishes a network of interdependence among farmers, promoting a sense of solidarity. Additionally, surplus produce can be traded in local markets, allowing farmers to generate income or obtain goods they cannot produce themselves.

Furthermore, bartering and trading surplus produce can act as a risk management strategy. In the face of unexpected challenges such as crop failures or adverse weather conditions, a diversified network of trading partners provides a safety net for farmers.

In conclusion, bartering and trading surplus produce are integral components of self-sufficiency farming. These practices not only contribute to the efficient allocation of resources but also strengthen community bonds, enhance food security, and provide a buffer against unforeseen challenges in the pursuit of sustainable and resilient agricultural practices.

3. Developing skills for self-reliance beyond farming.

- **Basic Survival Skills:** Learning to build shelters, start fires, find clean water sources, and forage for food are fundamental for self-reliance in any environment.

- **First Aid and Medical Skills:** Knowing how to administer basic first aid, treat injuries, and handle medical emergencies empowers individuals to take care of themselves and others in times of need.

- **Food Preservation and Cooking:** Understanding techniques for preserving food such as canning, drying, and fermentation, along with basic cooking skills, ensures access to nourishing meals even without access to modern conveniences.

- **Craftsmanship:** Proficiency in basic carpentry, sewing, and other crafting skills enables individuals to repair and create essential items, reducing dependence on external resources.

- Navigation and Orientation: Mastering navigation using maps, compasses, and natural landmarks allows individuals to travel safely and confidently, whether in urban or wilderness settings.

- **Self-defense:** Learning techniques for self-defense and situational awareness enhances personal safety and security, fostering independence and resilience.

- **Financial Management:** Developing skills in budgeting, saving, and investing ensures financial stability and independence, enabling individuals to weather economic uncertainties and pursue their goals.

- **Communication and Networking:** Cultivating effective communication skills and building strong social networks fosters collaboration, support, and access to resources within communities.

- **Problem-Solving and Critical Thinking:** Nurturing the ability to analyze situations, generate solutions, and adapt to challenges fosters resilience and innovation, key qualities for self-reliance.

- **Emotional Resilience and Mental Well-being:** Developing coping strategies, emotional intelligence, and practices for maintaining mental health equips individuals to navigate life's ups and downs with strength and grace.